IMAGES
of Aviation

FLOYD BENNETT
FIELD

FLOYD BENNETT FIELD TODAY. A passenger sitting in an airliner flying a Parkway Visual Runway 13L/R approach to Kennedy International Airport might notice what appears to be a complete airfield lying dormant on the western edge of Jamaica Bay. It is the former Naval Air Station New York, Floyd Bennett Field, now part of the Gateway National Recreation Area and the subject of this book. (Dmitry Avdeev.)

ON THE COVER: On September 2, 1936, entertainer Harry Richman and Eastern Airlines captain Dick Merrill attempted a nonstop flight to London in this Vultee V1-A, *Lady Peace*. Equipped with extra fuel tanks, special Wright Cyclone engines, and 41,000 ping-pong balls (for better flotation in case of ditching), they were forced down by bad weather in Wales. They still set the transatlantic speed record of 18 hours, 36 minutes. (Cradle of Aviation Museum.)

IMAGES
of Aviation

FLOYD BENNETT
FIELD

Richard V. Porcelli

ARCADIA
PUBLISHING

Published by Arcadia Publishing
Charleston, South Carolina

Printed in the United States of America

Library of Congress Control Number: 2015931286

For all general information, please contact Arcadia Publishing:
Telephone 843-853-2070
Fax 843-853-0044
E-mail sales@arcadiapublishing.com
For customer service and orders:
Toll-Free 1-888-313-2665

Visit us on the Internet at www.arcadiapublishing.com

To Linda, my wife and partner, without whom I would never be able to realize my dreams. And to the men and women of our armed forces, and their families, who selflessly sacrifice so much for our freedom and to preserve our way of life.

CONTENTS

ACKNOWLEDGMENTS

As is true for many projects that delve into history, the story of Floyd Bennett Field became more intriguing and remarkable the deeper I researched, with so many stories of success and failure, heroism and tragedy, and service to our country. This project has also given me the opportunity to meet and share a mutual love of aviation history with so many generous people who assisted in this project. I would like to particularly thank Capt. Dennis Irelan and Comdr. Doug Siegfried of the Tailhook Association for their continual help and support; J. Lincoln Hallowell, park ranger, and Felice Ciccione, archivist and curator, of the National Park Service/Gateway Recreation Area for their assistance as well as for their efforts to preserve and restore Floyd Bennett Field; Joshua Stoff, curator, and especially Julia Bloom, archivist of the fantastic Cradle of Aviation Museum in Garden City, Long Island, for the invaluable images of aviation's golden age; S.Sgt. Christopher Muncy of the 106th Rescue Squadron, New York Air National Guard for images of the little-known air national guard (ANG) activity in Brooklyn; fellow enthusiast and photographer Dmitry Avdeev for allowing me to use his aerial photography; the always helpful and patient staff of the National Archives, the source of so many important images; Robert Henshaw of the Naval History and Heritage Command; Jack Coyle of the National Naval Aviation Museum; as well as Dick Arkins and Roger Stites of the Vought Aircraft Heritage Center, all for allowing me to use priceless images from their archives. I also want to acknowledge the constant encouragement from my brothers, Joe and Bob, who triggered my passion for aviation in the first place.

Finally, how do I thank my wife, Linda, properly and keep within the word limit for this book? She has always supported, encouraged, and counseled me in all my endeavors. She has kept me on course throughout this and countless other projects. She has been an invaluable proofreader, organizer, fellow researcher, and sounding board, and as always, she remains my best friend.

Photographs are credited as follows:

CAM—Cradle of Aviation Museum
DA—Dmitry Avdeev
NA—National Archives
NH—Naval Historical Center/Naval History and Heritage Command
NNAM—National Naval Aviation Museum
NPS—National Park Service/Gateway NRA Museum Collection
NYANG—New York Air National Guard/106th Rescue Wing
TH—Tailhook Association
USCG—United States Coast Guard Historical Center
Vought—Vought Aircraft Heritage Center Archives

INTRODUCTION

After the World War I, US transportation remained dominated by the railroads and steamships. New York City, a major hub for both of these systems, was slow to embrace the importance of air travel. The government stimulated air commerce by promoting airmail service, and in 1927 the Post Office Department transferred airmail service to economically ailing airlines. Commercial airport construction blossomed, with the first (and the source of the term "airport") being in Atlantic City, New Jersey. New York City lagged behind despite the fact that significant flights, such as those of Lindbergh, Chamberlin, and Byrd, originated in the metropolitan area; instead, Newark, New Jersey, opened the area's first municipal airport in 1928. The Post Office Department responded by placing New York City's airmail terminus in New Jersey. This perceived insult and potentially significant loss of revenue stimulated a response from New York City officials.

With the airport at Newark under construction, New York City began its search for the site of its first municipal airport. A committee studied 10 alternative sites in six general regions, one being the Jamaica Bay area south of the boroughs of Queens and Brooklyn. The commission asked noted aviator Clarence Chamberlin to evaluate the alternative sites and then design the airfield. (Just a week after Lindbergh's famous May 1927 solo flight, Chamberlin flew a Bellanca nonstop from Roosevelt Field, Long Island, to outside of Berlin, Germany. Despite the greater distance, Chamberlin's flight remains overshadowed by Lindbergh's.) In 1928, Chamberlin recommended the Barren Island site on the western end of the Jamaica Bay region, while the majority of the other committee members preferred a site on the eastern end—one that coincidentally is the location of Kennedy International Airport. The committee, however, went with Chamberlin's recommendation.

What was called Barren Island was actually a peninsula surrounded by small islands and attached at its northern end to Brooklyn, 13 miles from Times Square. Once the home of pirates, rumrunners, and squatters, the small Brooklyn community had a main street, a few houses and shops, a church, and a post office; it was best known for its garbage dump and rendering plant. The adjacent glue factory, with its distinctive 200-foot-high chimney, was used for the disposal of dead horses, the primary means of transport in the city through the 1920s. The body of water just west of the site is known to this day as Dead Horse Bay.

The city allotted $5 million for building up and leveling of the land as well as connecting numerous islands with six million cubic yards of sand pumped from Jamaica Bay, reclaiming 321 acres raised 16 feet above the low-tide water level of Jamaica Bay. Construction work began October 29, 1929. Perhaps in an ominous coincidence or, as some say, a prediction of its disappointing future as a commercial airport, that was the same day of the stock market crash and the start of the Great Depression.

In a time when most airports had sod, dirt, or gravel runways, the new airport could boast two 100-foot-wide, reinforced concrete runways that intersected just south of an administration building. Special features included a floodlight system that allowed for night operations—a unique feature for the day—as well as a directional radio beam and special border lights to aid landing aircraft. A seaplane ramp was constructed on the opposite (east) side of the airfield.

Eight hangars were built in sets of two hangars each, parallel to Flatbush Avenue. Hangars 1–4 were located north, and Hangars 5–8 were south of a centrally located administration building. This red and black brick building uniquely served both as the passenger terminal as well as for airport offices. Sited east of Flatbush Avenue, the rectangular building had projections for the entry portico on the west side and the control tower on the east side. Some unique features of the building's design included ground-level entrance ramps from the street side into the basement level, which connected to baggage ramps on the "air side" of the building. Later, during an upgrading program financed by the Works Progress Administration (WPA), four underground tunnels leading to the individual airliner parking spots on the apron were installed. They allowed passengers to board aircraft safely regardless of the weather and out of the danger of spinning propellers. Four roundtables were set in the pavement to swivel airliners on what was hoped would be a congested ramp. The administration building also featured a barbershop, pilot's lounge, weather bureau office, restaurants, and observation deck. The interior was finished in the colorful Art Deco style, including French doors and a glass ceiling.

The airport was dedicated on June 26, 1930, to honor Brooklynite and Naval Aviator No. 9, Floyd Bennett. Bennett made history in 1926, when he flew with Adm. Richard E. Byrd on the first flight over the North Pole. The flight earned both Navy men the Medal of Honor, although years later, historians questioned whether they had in fact flown over the pole. In 1928, Byrd and Bennett were planning for a flight over the South Pole, but during a test flight, Bennett crashed, breaking several ribs and puncturing a lung. While recovering from injury-caused pneumonia, Bennett insisted on piloting a rescue flight to aid the crew of the Junkers W33 aircraft named *Bremen*, which crashed in northern Canada during the first east-to-west nonstop transatlantic flight. Bennett's condition worsened, and he died during the trip despite an effort by Charles Lindbergh to fly serum to the ailing aviator. Naming the new airport to honor a naval aviator was prescient of the major role that the Navy would play in the future of the airport.

New York's Metropolitan Airport No. 1 at Floyd Bennett Field officially opened on May 23, 1931. However, despite being the most advanced airport of the time, it never achieved commercial success. It suffered from a number of weaknesses, especially its location in the far reaches of Brooklyn, only accessible via congested Flatbush. A seaplane service to Manhattan failed to resolve the problem. Passengers and, most importantly, the Post Office Department argued that Newark Airport was far more convenient, a position bolstered by the 1932 opening of the Pulaski Skyway that linked Newark Airport to the Holland Tunnel and Manhattan. During the Depression, with more than 25 percent unemployment and average salaries less than $1,400 per year, there were few people that could afford air travel—hence the critical importance to airlines of airmail service revenue.

In a famous November 1933 publicity stunt, Fiorello LaGuardia bought a ticket on Transcontinental and Western Airlines from Chicago to New York, but after the aircraft arrived at Newark Airport, LaGuardia refused to deplane, arguing that his ticket said the destination was "New York" and demanding to be flown to New York City. The airline, fearing a lawsuit and adverse publicity, complied, and LaGuardia gave an impromptu press conference to reporters on the short hop to Floyd Bennett Field, urging New Yorkers and Washington bureaucrats to support the New York City airport. But, the efforts were to no avail, and in March 1936, the postmaster general made the final decision, refusing to move New York City's mail service away from Newark Airport, citing superior accessibility to Manhattan and the railroads and thus sealing the commercial fate of the new airport. The final blow was the opening of New York's Municipal Airport No. 2 (today's LaGuardia Airport) at North Beach, Queens, which benefitted from a closer location and access to the city via new highways, bridges, and the subway.

The Navy's connection with Floyd Bennett Field actually began years before its opening. The Navy's first presence was a balloon base built during World War I at Fort Hamilton, Brooklyn. It moved to Rockaway Beach, an airship and seaplane base built during World War I for coastal protection, just across Jamaica Inlet from Barren Island. A Naval Reserve air base (NRAB) was established there in 1923 after the active Navy left. The Reserves stayed until forced to move by the city in 1929. Wanting to relocate to the new Floyd Bennett Field, the NRAB spent a year at the Curtiss-Wright Airport in nearby Valley Stream. It became the first "tenant" of Floyd Bennett Field in 1931, moving into offices in the administration building plus Hangar 5. At Floyd Bennett Field, the NRAB provided "elimination training" of aspiring aviators in an effort to select the best pilots for basic training at Pensacola. It also provided flying time for the many Navy reservists who lived in the New York area.

Another early tenant, based in Hangar 3, was the NYPD's Aviation Unit. This, the country's first police aviation department, was formed in 1919 with two seaplanes donated by the Navy under the proviso that the police pilots join the Navy Reserve. The US Coast Guard leased space on Jamaica Bay starting in 1936 and in 1938 opened one of its first 10 air stations at Floyd Bennett Field. Air Station Brooklyn, as it was called, grew and continued to operate from the same site until it was closed in 1998.

The only scheduled airline service from the Floyd Bennett Field was short lived. In 1937, American Airlines began service linking New York and Boston, which ended in 1938. Although Floyd Bennett Field was never a commercial success, it was the second-busiest airport in the country in 1933, with almost 52,000 operations, mainly private pilots, and, historically more significant, record-breaking attempts. Its appeal was based on its modern facilities (especially long, paved runways), landing aids, lack of nearby obstacles, generally good weather (compared to other regional airports), and a location as a perfect jump-off point for transatlantic or transcontinental flights. It became a favorite airfield of many aviators, especially those striving to set speed and distance records during this "Golden Age of Aviation." The list of successful (and failed) attempts as well as resulting technical achievements would fill pages. From 1931 to 1939, some 26 "round-the-world" and transoceanic flights, as well as 10 transcontinental record-breaking flights, began or ended at Floyd Bennett Field. Notable aviators included James Haizlip, James Doolittle, Wiley Post, Harold Gatty, Amelia Earhart, James Mattern, Bennett Griffen, George Pond, Cesare Sabelli, Russell Boardman, John Polando, Clyde "Upside-Down" Pangborn, Hugh Herndon, Dick Merrill, Harry Richman, Roscoe Turner, Jacqueline Cochran, Alexander de Seversky, Howard Hughes, and Douglas "Wrong Way" Corrigan.

In 1939, as the fears of world conflict grew, the record-breaking flights ceased while the Navy bolstered its facilities. The NRAB expanded, taking over Hangar 2. The active Navy leased additional acreage to build a new seaplane base, Hangar A, adjacent to the Coast Guard Air Station to be used for squadrons assigned to the Neutrality Patrol. Through 1941, the tempo of Navy activities grew; the Reserve units were called to active duty, and new facilities were constructed. In May 1941, the Navy leased the entire airfield from New York City for $1 per year, evicted the last residents of the Barren Island community, stopped all nonmilitary flying, and established Naval Air Station New York (NASNY). In January 1942, the Navy relieved New York City of this financial burden, paying $9.25 for the airfield. Additional facilities were built, including barracks, workshops, hangars, seaplane ramps, and a wharf. Although one of the original runways was closed because of its proximity to the administration building, the remaining ones were widened and lengthened, and another runway was added.

With the buildup of warplane production authorized by Congress, the need for an efficient way of getting new aircraft from the factory to combat units became apparent. NASNY was centrally located amidst many of the naval-aircraft production factories. Chance Vought and Vought Sikorsky (Stratford, Connecticut), Grumman (Bethpage, New York), GM-Eastern Aircraft (Linden and Trenton, New Jersey), Brewster (Newark, New Jersey), Hall Aluminum Aircraft (Bristol, Pennsylvania) and Martin (Baltimore, Maryland) factories were within 150 miles of Floyd Bennett Field. An aircraft-delivery unit was established at NAS New York in August 1941 to ferry in new aircraft manufactured by factories in the area, to equip them completely for combat action in the fleet, and to ferry them out to their operating units. This service was expanded with the establishment of the Naval Air Ferry Command, headquartered at NASNY, which occupied Hangars 9 and 10 that were added to Hangar Row, and Hangar B built on the east side of the airfield. More than 20,000 new aircraft were accepted and ferried from Floyd Bennett Field. Taking into account the ferrying of "war weary" aircraft back from combat units to the factories and repair facilities for refurbishment, 46,000 aircraft movements were recorded at NASNY between December 1943 and November 1945. The resident Naval Air Ferry Command was responsible for more than 75,000 aircraft delivered during that period.

NASNY also became a terminus for the Naval Air Transport Service (NATS). As the war progressed, so did the NATS operations, with two squadrons providing both transcontinental and transatlantic passenger and cargo service.

During the war years, the Coast Guard and active Navy squadrons flew countless antisubmarine and air-sea rescue missions from Floyd Bennett Field. More significantly, in 1943, the Coast Guard established the country's first and only helicopter training base. Sikorsky R-4/HNS-1 Hoverflys, the first production helicopter, were used to train hundreds of pilots and mechanics for all branches of the US and British military forces. The first of literally hundreds of thousands of life-saving missions attributed to the helicopter was in January 1944, when blood plasma was flown by helicopter from Brooklyn to Sandy Hook, New Jersey, for sailors rescued from the explosion and sinking of USS *Turner* in lower New York harbor.

After the end of World War II, active-duty Navy operations dropped drastically, but NASNY became the largest Naval Air Reserve base in the country. At its peak, the air station housed 34 Navy and Marine Reserve squadrons, supporting the activities of more than 3,000 "weekend-warrior" reservists. In February 1951, five Navy and Marine Reserve squadrons were recalled for active duty as a result of the outbreak of the Korean War. Fighting Squadrons 831 and 837, crewed entirely by Brooklyn-based reservists, flew combat missions from USS *Antietam* from October 1951 to March 1952. James Michener's *The Bridges at Toko-Ri* is based in part on interviews with the reservists from Floyd Bennett Field. Later, during the Cold War, NASNY's Reserve units were also recalled for active duty in response to the Berlin Crisis of 1961 and the Cuban Missile Crisis of 1962. The postwar years also saw the resumption of transcontinental record-breaking flights by Navy aircraft, including ones flown in 1957 by Marine pilot and then NASA astronaut Sen. John Glenn and in 1961 by Navy pilot then NASA astronaut Richard Conrad.

Going back to 1947, Floyd Bennett Field also became the home of New York Air National Guard's 106th Wing, the oldest unit in the Air National Guard. Originally comprised of two light bomb squadrons, the unit was also mobilized in February 1951, converted to B-29 Superfortresses, and transferred to March Field, California, to train replacement bomber crews for Korean War operations. They were released from active duty during 1953; returning to Floyd Bennett Field, they went through a progression of equipment, designation, and role changes, including bombardment, fighter interception, medical evacuation, heavy transport, and finally air-to-air refueling. In 1970, the Air National Guard moved 75 miles east to the recently vacated Suffolk County Air Base, now called the Francis Gabreski Air Guard Base, where the 106th Air Rescue Wing flies Sikorsky Pave Hawk helicopters and Lockheed Hercules cargo/tanker planes as one of the Air National Guard's three combat search-and-rescue wings.

By 1970, the Navy was deep into cost cutting because of the heavy financial burden of the Vietnam War. With the resulting drastic reorganization of the Naval Air Reserve, the end of NASNY was in sight. The naval air station was closed in mid-1971, leaving only the NYPD and USCG to operate out of the west side Hangar 1 and east side Coast Guard Air Station, respectively. The airfield itself was turned over to the Department of the Interior.

Since 1972, Floyd Bennett Field has been part of the National Park Service's Gateway National Recreation Area. Many of the original structures that have survived have been listed in the National Register of Historic Places. Fortunately, the park rangers responsible for Floyd Bennett Field also recognize its military and specifically naval-aviation historical value. The National Park Service, in conjunction with an avid group of aviation historians and Navy retirees known as the Historic Aircraft Restoration Project (HARP), occupies Hangar B, where it restores and maintains a collection of historic aircraft representative of the history of Floyd Bennett Field. Today, the Coast Guard is gone (to Air Station Atlantic City), and its former facility is now home of the NYPD, the last remaining flying unit based at Floyd Bennett Field.

Under the National Park Service, Floyd Bennett Field has also become a bird sanctuary, the largest urban campsite in the country, a site for unrestricted star gazing (based on its distance from light pollution), and a place for public vegetable gardens, bicycling, fishing, and many other activities that this unique location offers to otherwise apartment-living urban dwellers.

The historical significance of Floyd Bennett Field is compelling. It was New York City's first municipal airport, and it was the most modern airport of its day. Although not a commercial success, it was the site of so many events that advanced aviation technology during aviation's golden age. Naval Air Station New York at Floyd Bennett Field played a significant role in World War II, including in the Coast Guard's development of the use of helicopters in their life-saving role. After the war, it became the country's largest Naval Air Reserve base, and Navy and Air National Guard reservists served during the Korean War and subsequent Cold War crises. Perhaps most importantly, Floyd Bennett Field is probably the only example of a 1930s-era airport anywhere in the world that remains intact, virtually as it was originally built.

One

THE ROARING TWENTIES BRING HIGH ASPIRATIONS

NAVY KITE BALLOON AT ROCKAWAY, AUGUST 1918. The story of Floyd Bennett Field cannot be told without describing the persistent influence of naval aviation on the course of the airfield's development. NAS Rockaway Beach was constructed in 1917 as part of a chain of coast patrol stations. Coincidentally, its location was across the inlet from the future Floyd Bennett Field (NNAM.)

NAS Rockaway Beach, December 1918. NAS Rockaway Beach was built in 1917 on property transferred from New York City's Borough of Queens. Seen from the air, the extensive facilities of the air station, located on the inlet side of Rockaway Peninsula, can be clearly seen. The large structure in the background above, looking east, is the main airship hangar. It is the leftmost structure in the photograph below. The gasometer, where hydrogen was stored, is adjacent to the hangar. The smaller hangars along the shore were for Navy flying boats and floatplanes. Although there were no proper runways, land-based planes could use the clear area just south of the airship hangar. (Both, NA.)

AIRSHIPS OVER ROCKAWAY BEACH, 1918.
A B-class airship is shown flying over a
tethered kite balloon. The Navy purchased 20
of these airships for service during and shortly
after World War I. Most were assigned to East
Coast air stations for antisubmarine duty. The
gondola, slung beneath the gasbag, was actually
a modified JN-4 Jenny fuselage; power was from
a 100-horsepower OX-5 engine. (NNAM.)

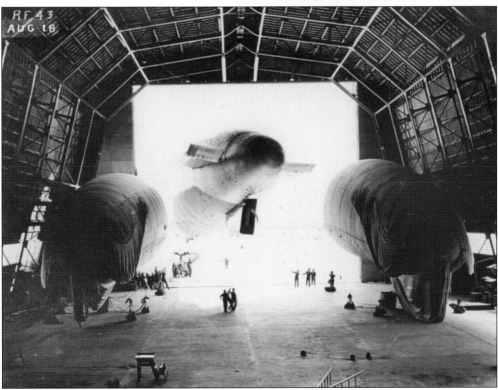

AIRSHIP HANGAR, 1918. Built during the American involvement in World War I, the air station
boasted a state-of-the-art airship hangar. Here, a B-class airship is being guided out of the hangar
that also contains two kite balloons. Manned kite balloons were flown tethered to ships or the
ground for observation purposes. (NNAM.)

PARACHUTE DEVELOPMENT, 1919. NAS Rockaway was involved in experimentation as well as offshore patrols. One experiment was with a primitive form of parachute harness, modeled here by a naval officer. Since tethered observation balloons were tempting targets for enemy aircraft, the observers needed a way to escape if attacked. (NA.)

DIRIGIBLE AT NAS ROCKAWAY BEACH, 1919. The C-type airship shown here was a very successful design for coastal patrol and convoy escort duty. It greatly influenced the development of nonrigid airships by both the Navy and the Army. The crew of four sat in the boat-like car that was equipped with a pusher motor mounted on each side. (NA.)

CURTISS MF FLYING BOAT, 1918. The Curtiss MF is sitting on a ramp outside one of the smaller hangars. The wooden-hulled MF was an improved flying boat based on the venerable F model. A 100-horsepower OXX engine drove the pusher-type propeller. Curtiss built 22 examples at their nearby Garden City, Long Island, factory. (NA.)

CURTISS HS FLYING BOAT, OCTOBER 1918. A Curtiss HS is shown taxiing on the Jamaica Inlet. The smokestack from the glue factory across the inlet identifies the site of the future Floyd Bennett Field. During World War I, Curtiss and subcontractors built 1,178 examples—an enormous number for the times. The 360-horsepower V-12 Liberty engine gave it a top speed of 82 miles per hour. (NNAM.)

Curtiss H-16 Being Assembled, October 1918. The Curtiss Aeroplane Company factory building, located in Garden City, Long Island, was not large enough to hold fully assembled aircraft the size of their H-16 flying boats; therefore, components were shipped to NAS Rockaway Beach for final assembly and flight-testing. (NA.)

Curtiss H-16 at NAS Rockaway Beach, 1919. The flying boat is sitting at the foot of a ramp leading from one of the seaplane hangars to the water; it is on a trolley used to ease movement of the aircraft. Barren Island, where Floyd Bennett Field would be built, is across the inlet. The H-16 flying boats were large for their day, with a 95-foot wingspan. (NA.)

CURTISS F-5LS AT NAS ROCKAWAY, 1919. These flying boats were derived from the Curtiss Model H built in 1914 under special order by Rodman Wanamaker for a transatlantic attempt. Sold to the Royal Navy, it became the basis of the Felixstowe F series of aircraft produced in Britain. The design was then imported back to the United States, improved upon, and produced by Curtiss Aeroplane Company (in Garden City, Long Island, and Buffalo, New York) as well as by the Naval Aircraft Factory in Philadelphia as the H-16. The flying boat cruised at 95 miles per hour and was powered by two 400-horsepower Liberty V-12 engines. (Both, NA.)

FLYING BOATS IN FLIGHT, 1919. Two Curtiss HS flying boats are shown near NAS Rockaway Beach. The Curtiss HS in the background shows squadron insignia of the period. NAS Rockaway Beach, in the borough of Queens, is just below the trailing aircraft, while the leading aircraft is over neighboring Brooklyn. Note the front gunner's position is fully exposed to the air. (NNAM.)

RARE FLOATPLANE AT NAS ROCKAWAY BEACH, 1919. The Curtiss HA Dunkirk fighter was designed as a two-seat escort and air-superiority fighter for use during World War I over the Dunkirk-Calais area—hence its name. Powered by a 400-horsepower Liberty engine, the prototype crashed; two further examples were built. Both were used for postwar testing at NAS Rockaway Beach. (NNAM.)

NAVY-CURTISS (NC) FLYING BOATS, 1919. Starting out as a design for U-boat hunting seaplanes during World War I, the NC flying boat was then selected for the Navy's attempt of the first flight across the Atlantic Ocean, albeit in stages. Four aircraft, NC-1 to NC-4, were built for the flight by Curtiss at its Garden City factory but were assembled and test flown at NAS Rockaway Beach. The photograph above shows the ongoing assembly of one of the aircraft, while the one below shows a completed NC-2 undergoing tests. Equipped with three forward-facing tractor engines and a fourth center-mounted, rear-facing pusher engine, the flying boat had a crew of six. (Above, NA; below, NNAM.)

INSPECTION PRIOR TO DEPARTURE. On May 8, 1919, three NCs (NC-2 proved unreliable in testing and was cannibalized for spare parts) departed Rockaway Beach. Only NC-4 completed the record-breaking flight, arriving in Lisbon, Portugal, on May 31. Inspecting NC-3 are, from left to right, Capt. John Towers, flight commander; Capt. Noble Irwin, director of aviation for the Navy; and Comdr. Holden Richardson, NC-3 pilot. (NA.)

UO-1C AT NAVAL RESERVE AIR STATION NEW YORK, 1928. Vought built 141 UO observation planes at its nearby Long Island City factory. In early 1922, after suffering the effects of a number of hurricanes and a calamitous hydrogen fire that destroyed the airship hangar, NAS Rockaway Beach was disestablished; however, in August 1923, the Naval Reserve air station originally established at Fort Hamilton, Brooklyn, moved to Rockaway Beach. (NNAM.)

CURTISS-WRIGHT AIRPORT. In 1929, the City of New York notified the Navy that it wanted the Rockaway Beach site back for reversion to its original recreational use, forcing the Naval Air Reserve Unit (NARU) in January 1930 to move 10 miles east to Curtiss-Wright airport in Valley Stream, Long Island, shown here. Today, the hangars still exist, but the runways are buried under a shopping center. (NA.)

NARU CURTISS F4C-1. Originally designated as the TS-1, this diminutive biplane fighter was noteworthy as being the first carrier-based aircraft specifically designed for that purpose. It flew from the first aircraft carrier, the USS *Langley*. After retirement from active duty units, NARUs, including this one, were equipped in Valley Stream. (NA.)

NARU AMPHIBIANS. In the photograph above, a NARU Loening OL amphibian is flying over the Brooklyn Bridge, which connects Manhattan with Brooklyn. Grover Loening, a Columbia University School of Engineering graduate, designed and built numerous amphibians (seaplanes with extendible landing gear) for the Navy until Grumman Aircraft acquired his company. While at Rockaway Beach, the Naval Air Reserve Unit flew both land and seaplanes, but since the inland Curtiss-Wright Airport in Valley Stream could only accommodate the former, an additional base for the seaplanes was needed. They ended up at the Curtiss-Wright Air Terminal in North Beach, Queens, pictured below. Coincidentally, that airport would become Municipal Airport No. 2, (today's LaGuardia Airport) and play a role in the ultimate demise of the commercial aspirations of Floyd Bennett Field. (Both, NA.)

Two

COMMERCIAL DISAPPOINTMENT BUT INCREASING NAVY PRESENCE

WPA FEDERAL ART PROJECT POSTER, 1936. New York City mayor Fiorello LaGuardia was a staunch advocate of aviation and fought hard to establish commercially viable municipal airports for his city. His plans for the first, at Floyd Bennett Field, were frustrated by the Post Office Department's unwillingness to move from Newark Airport. The second, at North Beach, was eminently successful and is known today as LaGuardia Airport. (NPS.)

NEWARK METROPOLITAN AIRPORT, MAY 1929. Newark, New Jersey, built its metropolitan airport three years earlier than New York City and secured the vitally important airmail distribution center; however, as shown in this photograph, it was a very rudimentary airfield with a sod runway. In contrast, Floyd Bennett Field was built as the most modern airfield in the world, with two concrete runways and modern landing aids. (NA.)

FLOYD BENNETT FIELD (FBF), SEPTEMBER 1931. While its location was selected partly due to favorable weather trends, its remote location at the far reaches of Brooklyn was one of its fatal flaws. This photograph shows the layout of the two runways that crossed just east of the administration building. Four pairs of freestanding hangars flank the building, while Flatbush Avenue runs across the photograph. (NA.)

US Navy Machinist Floyd Bennett.
The new airfield was named after
Brooklyn resident and Medal of Honor
winner Floyd Bennett. Bennett (at
right) enlisted in the Navy in 1917
for flight training and later met
Adm. Richard E. Byrd. Impressed
with his flying skills, Byrd chose
Bennett as his pilot for an aviation
survey of Greenland in 1925 and for
an attempt to reach the North Pole.
In May 1926, flying a Fokker BA-1
named *Josephine Ford* (below), they
took off from Spitsbergen, Norway,
for what was believed to be the first
flight over the pole. Byrd and Bennett
were awarded the Medal of Honor
for their effort, but historians later
questioned whether they had in fact
reached their objective. (Both, NH.)

ADMINISTRATION BUILDING. The administration building was described as an eclectic mix of elements, with influences ranging from the Renaissance Revival, Colonial Georgian Revival, Neoclassical, and Art Deco styles. This hybrid of styles was in fashion in the early 20th century for public structures in New York City and elsewhere. The west, or "land," side (above) contained an entrance portico for passengers arriving by automobile. Their luggage would be sent through separate street-level doors to the lower floor. In 1935, as part of a terminal-improvement program financed by the Works Progress Administration, underground walkways were added on the east, or "air," side (below) that allowed passengers to walk directly to the door of their aircraft, avoiding bad weather and spinning propellers. (Both, NPS.)

PERSPECTIVE OF ADMINISTRATION BUILDING AND HANGAR ROW. The photograph above shows the central administration building, oriented along a north-south axis (parallel to Flatbush Avenue, seen behind the building) and one of the four pairs of brick hangars. Hangars 1 through 4 were on the south side, and Hangars 5 through 8 were on the north side. The four rectangular entrances to the underground walkways can be seen, as can four turntables used to swivel aircraft and avoid excessive taxiing on the cramped apron. These features are just some of the many advances incorporated in the airport design. Hangars 3 and 4 can be seen in the image below, which looks south from the roof of the administration building, with the aircraft-parking apron to the left. (Both, NPS.)

DOUGLAS DC-3S, 1937. The Douglas DC-3 truly revolutionized air transport in the 1930s. Seating 21 passengers in comfort, they could fly higher and faster than any of their predecessors or contemporaries. There were 16,079 civilian and military examples built, and a number are still flying. Although this photograph shows a lineup of American and Colonial Airlines' DC-3s, actual airline service was limited and disappointing. (NPS.)

WILEY POST GREETED BY 50,000 ADMIRERS. The lack of airline service was made up for by many record-breaking flights. Wiley Post completed the first solo round-the-world flight on July 22, 1933. Flying time in his Lockheed Vega 5C *Winnie Mae* was seven days, 18 hours, and 45 minutes, starting and ending at FBF. His solo flight relied on an autopilot and radio directional finder that Post was instrumental in developing. (CAM.)

LOCKHEED VEGA CENTURY OF PROGRESS. In June 1931, Wiley Post and navigator Harold Gatty successfully flew around the world from Roosevelt Field, Long Island. On July 5, 1932, James Mattern and Bennett Griffen set out from FBF in this red, white, and blue Vega in a failed attempt to beat the Post/Gatty round-the-world record. The aircraft crashed in Russia; fortunately, the crew survived. (CAM.)

ILL-FATED BELLANCA LITUANICA. Lithuanian American pilots Steponas Darius and Stasys Girénas departed FBF in their red-painted Bellanca on July 15, 1933, with their goal being Kaunas, Lithuania. They successfully crossed the Atlantic and actually flew 3,984 miles nonstop, only to crash for unknown causes near Soldin, Germany, 636 miles short of their goal. Both pilots perished. (CAM.)

BELLANCA PACEMAKER *LEONARDO DA VINCI*. On May 14, 1934, US Navy Reserve pilot George Pond (left) and copilot Cesare Sabelli flew this Bellanca Pacemaker from FBF in an attempted nonstop flight to Rome. Engine failure caused them to land in Ireland, frustrating a planned greeting by Benito Mussolini. Fortunately, they survived a crash of their Bellanca in Wales on their return trip. (CAM.)

BELLANCA PACEMAKER *ROSE MARIE*. Italian immigrant Giuseppe Bellanca founded the Bellanca Aircraft Company—a firm that still exists. In the 1920s and 1930s, he built many aircraft for pilots intent on breaking records (a Bellanca was Lindbergh's first choice). On June 3, 1932, Stanislaus Hausner attempted a transatlantic flight in this Bellanca CH to Warsaw but was forced to ditch at sea. (CAM.)

BELLANCA J300 LONG-DISTANCE SPECIAL, 1931. Clyde "Up-side Down" Pangborn and Hugh Herndon tried to break the record for circumnavigation of the world. Departing from FBF on July 28 in their red Bellanca named *Miss Veedol*, they were forced to give up their attempt due to poor weather conditions over Siberia. After repairs, they would later go on to complete the first nonstop transpacific flight. (CAM.)

SAVOIA MARCHETTI S.55. In July 1933, Italian general Italo Balbo led a transatlantic formation of 24 S.55 flying boats to Chicago for the Century of Progress Exhibition. On their return, they landed at FBF for a New York City visit. Although Balbo was a fascist, he disagreed with Mussolini's policies. In 1940, he was killed when Italian army gunners "accidentally" shot him down over Tobruk. (CAM.)

FLIGHT LINE, 1933. Although New York City's attempt to gain airline service at FBF was a failure, by 1933 the airport was the busiest in the country in terms of aircraft movements. It was very common for average New Yorkers to spend the day standing by the fence to watch private aircraft, such as the three Ryan B-1 Broughams and a biplane shown here, depart and arrive. (CAM.)

WEDDELL-WILLIAMS MODEL 44 RACING PLANE. Four examples of this aircraft were built in the 1930s; they were successful racing and record-breaking planes. Jimmy Weddell flew this aircraft, named *Miss Patterson*, to win the Bendix and Thompson Trophies. James Haizlip, shown here, set the Burbank-to-FBF speed record in 1932. Roscoe Turner also flew No. 44 to numerous victories, but he crashed in 1936 during a cross-country record attempt. (CAM.)

SEVERSKY IN SEV-AP-7A, 1938. Alexander de Seversky was the Russian military attaché in Washington, DC, when he gained asylum after the Russian Revolution. He became a major in the US Army Reserve and established his own aircraft company. He and others, notably aviatrix Jacqueline Cochran, set a number of records flying from FBF in aircraft, such as this SEV-AP-7A, built by his company. (CAM.)

RECORD-BREAKING LOCKHEED ELECTRA. In May 1937, Richard Merrill and John Lambie set records for FBF-to-Croydon (London) and Croydon-to-FBF flights with this Electra Model 10. On their return leg, they carried photographs of the coronation of King George VI for publication the next day in their sponsor's (Hearst Publishing) newspapers. (CAM.)

HUGHES BREAKS RECORD, 1938.
Eccentric industrialist and philanthropist Howard Hughes had a passion for aviation. During the 1930s, he established his own aircraft company, broke many world flying records, and purchased Trans World Airlines. In July 1938, Hughes piloted this Lockheed Super Electra, *New York World's Fair 1939*, to set a new round-the-world (RTW) record of 91 hours and 17 minutes. He and his crew of four were greeted at FBF by more than 25,000 people, at left, and then honored by a ticker-tape parade in New York City. The photograph below shows, from left to right, Mayor LaGuardia, official NYC greeter Grover Whalen, Hughes, and manager Albert Lodwick. (Both, CAM.)

Nazi Condor Lands at FBF. On August 11, 1938, Nazi pilot Alfred Henke and a crew of three flew a prototype Focke-Wulf FW-200 Condor, *Brandenberg*, from Berlin to FBF in a record-breaking 24 hours, 50 minutes. They also broke the eastbound record on their return flight. It was the only time a swastika-bearing German heavier-than-air aircraft landed at a US airport. (CAM.)

American DC-3, 1938. As a commercial airport, FBF was a disappointment. Despite being the nation's most advanced airport, it was only able to secure one daily commercial airline service. American Airlines began daily service to Boston in 1937 but discontinued it in 1938. (CAM.)

Consolidated NY-3s of NRAB New York. Consolidated Aircraft in Buffalo, New York, built the NY series of aircraft in the late 1920s and early 1930s. A 220-horsepower Wright engine powered the two-seat, open-cockpit trainer. The Navy acquired a number of these biplanes for elimination and primary flight training, with the NY-3 being used exclusively by the Naval Reserve air bases (NRABs). In the early 1930s, NRAB-NY used NY-3s to evaluate (and eliminate unqualified) aspiring naval aviators prior to being sent to Pensacola for primary flight training. In the photograph above, a NY-3 sits in the rain outside Hangar 1, while in the image below another example with the NRAB-NY emblem basks in the Brooklyn sunlight in 1936. (Both, TH.)

CURTISS AIRCRAFT OF **NRAB** NEW YORK. In the late 1920s and early 1930s, the Navy and Marines purchased F8C fighters and O2C observation planes from Curtiss that were derived from the Army Air Corps' Falcon fighter. Despite their designations, the Navy and Marines used them to develop dive-bombing techniques. Originally retaining the Army name, because of their new role both the F8C and O2C were called "Helldivers"—a very popular name Curtiss would reuse for a number of its later Navy aircraft. By 1934, both versions had been relegated to the Naval Air Reserve. In the photograph above, a NRAB-NY F8C-4 is shown in front of Hangar 7. The similar appearance to a Reserve O2C-1 taxiing down the ramp in the photograph below is obvious. (Both, TH.)

NRAB-NY Curtiss Helldivers, c. 1937. The Curtiss O2C-1 was built in Buffalo, New York. The crew, comprised of a pilot and observer/gunner, sat in open cockpits. The 450-horsepower Pratt & Whitney R-1340-4 engine gave it a top speed of 146 miles per hour and a service ceiling of 16,250 feet. It was one of the main types equipping the Reserve squadrons, including those at FBF. In the photograph above, the 1930s insignia of NRAB-NY is seen to good effect; it shows Mickey Mouse riding on a trident and bomb-laden goose past the Statue of Liberty. The view below of a sister ship gives a good view of the slightly swept outer wing panels. (Both, TH.)

MARTIN BM-1s, 1934. The development of dive-bombing techniques by the Navy and Marines, adding to the existing torpedo bombing skills, led to the evolution of more specialized aircraft types. The Martin BM series of dive-bombers was built in a Baltimore factory starting in 1931. The aircraft served with both fleet and Reserve squadrons, with some still flying in 1940. Shown above is a BM-1 (the second aircraft of Bombing Squadron Three, hence the code 3-B-2) visiting FBF for fleet exercises. Below, a NRAB-NY BM-1 is seen in a line with visiting aircraft, including a Martin T4M of Torpedo Squadron Five. (Both, TH.)

MARTIN T4M OF NRAB-NY, 1934. As part of the trend towards specialized aircraft, Martin also produced the T3M and T4M torpedo bombers for the Navy at its Cleveland, Ohio, factory. They became the standard carrier-based fleet torpedo bombers during the 1930s and began equipping the Reserves as well, starting in 1932. This T4M-1 on the FBF ramp proudly displays the NRAB-NY insignia. (TH.)

DOUGLAS XFD-1 AT NRAB-NY, 1934. The XFD-1 was an experimental fighter developed by Douglas Aircraft in response to a 1932 Navy request for a carrier-based biplane, competing with similar offerings from Vought and Curtiss. Testing was done at NAS Anacostia and FBF, as evidenced by the photograph of the only prototype in front of Hangar 5. The Navy decided against buying any of the contenders. (TH.)

VOUGHT CORSAIRS C. 1935. The Chance Vought Company continued its tradition of building Corsair-named scout bombers and observation biplanes for the Navy after it moved from Long Island to a new factory in Connecticut. Built as O3U observation planes, they were redesignated SUs to reflect their primary scout-bombing mission. During production, the tail was enlarged and reshaped. Above, a SU-3 of Bombing Squadron Five is on the FBF ramp. Hangar Row and the administration building are in the background. The photograph below shows a later SU-3 (with the larger tail) of the fleet's Battle Force parked nearby. (Both, TH.)

NRAB INSPECTION, 1939. The inspection of aircraft of NRAB-NY shows a lineup of Grumman SF-1s and J2Fs facing Hangar Row and the administration building. The SF-1 was a scout version of the FF-1 fighter, the first of many Navy aircraft built by the "Grumman Iron Works." It was the first Navy aircraft with retractable undercarriage and enclosed cockpit. The J2F Duck amphibian served the Navy and Coast Guard from 1934 until 1945. (TH.)

CURTISS SBC-4, 1940. The Curtiss SBC family of dive-bombers was also given the Helldiver name. It was destined to be the last combat biplane produced in the United States. Production started in 1934, and almost 200 remained in service at the start of World War II. This NRAB-NY Helldiver shows off the new 1940 insignia depicting Father Knickerbocker, a fictional symbol of a native New Yorker. (TH.)

NAVAL AIR RESERVE STUDENTS, 1941. NRAB-NY's main tasks were the elimination training of prospective Reserve naval aviators and the maintenance of flying skills of the existing reservists. In the image above, students push yellow-painted Naval Aircraft Factory N3N-1 trainers (known as "Yellow Perils") out of a FBF hangar in preparation for the day's flying. Below, flight students await their turns to man one of the training aircraft flown by the unit. Instructors, sitting in the rear cockpit of the two-seat training aircraft, would monitor the progress of the fledgling aviators and determine if they were suitable to continue primary flight training at Pensacola. (Both, NNAM.)

COAST GUARD AIR STATION BROOKLYN. The US Coast Guard had already leased land on FBF from the city in 1936 for use as a lifesaving station. In April 1938, it established on that site one of 10 air stations authorized by Congress. A large hangar, shown under construction in the photograph above, was built on the eastern edge of the airfield. Seaplane ramps and parking aprons were also built, as seen below. The unit was equipped with Hall-Aluminum PH-2 flying boats and Grumman JF amphibians. Taxiways linking the Coast Guard ramp to the runways were not added until later. (Above, NNAM; below NA.)

USCG RD-4 Dolphin, 1935. The Coast Guard ordered 13 Douglas RD flying boats in four different models. It featured an all-metal hull with seating for eight crewmen, a plywood-covered cantilever wing, and two 450-horsepower Pratt & Whitney R-1340-96 engines. Here, RD-4 No. 131 from CGAS Brooklyn named *Mizar* is sitting at the East River seaplane base. (USCG.)

Grumman JF-2 Steam Cleaning, 1937. Coast Guardsmen have removed the cowling and are steam cleaning the engine of this Grumman JF-2 Duck amphibian. Comparison with photographs of the earlier Loening OL amphibian shows the strong similarity, perhaps due to the fact that Leroy Grumman was a Loening employee prior to founding his own firm. (NA.)

GRUMMAN JF-2 DUCKS, C. 1939. The XJFG-1 prototype first flew in 1933. While the Navy was the main customer of the 570 produced by Grumman (and subcontractor Columbia Aircraft), in 1934 the Coast Guard acquired 14 JF-2s with special equipment and 750-horsepower Wright Cyclone engines. They set records and demonstrated air operations from its cutters. These photographs show Coast Guard JF-2s outside Hangar 1 (above) and in front of the administration building (below). The unique appearance of the brick hangars and administration building makes the Floyd Bennett Field location easily recognizable. (Above, NA; below, TH.)

Unique Northrop RT-1, March 1939. The Coast Guard bought a single Northrop RT-1 Delta, named the *Golden Goose*, in 1935 as the executive transport for the commandant as well as the secretary of the treasury. It is shown here visiting FBF just prior to being damaged in an accident that ended its service. (TH.)

Air Station Brooklyn Ramp, 1938. This view of the Coast Guard ramp at FBF shows a Stinson RQ-1 (left) and a Hall PH-2 flying boat. In 1934, one Stinson was purchased expressly for electronics testing at FBF; it was transferred to Cape May, New Jersey, in 1939 for coastal patrol. The PH-2 (right) was one of seven purchased by the USCG. (TH.)

USCG Hall Flying Boats in Flight. The Hall Aluminum Company of Bristol, Pennsylvania, built seven each of the PH-2 and PH-3 versions of these flying boats for the Coast Guard. A PH-2 is shown above over the shoreline near FBF, while another, seen below, overflies the Statue of Liberty and the departing Cunard liner RMS *Queen Mary*. While initially used for search-and-rescue duty, as war clouds loomed they conducted antisubmarine patrols as well. For that duty, their bright, prewar, polished-aluminum color scheme, with red, white, and blue tail markings, was sprayed over with dull gray paint. (Both, TH.)

Three

CIVIL AIRFIELD TO WARTIME NAVAL AIR STATION

OVERALL PERSPECTIVE JUST BEFORE THE WAR. This photograph, looking east, shows the layout of FBF towards the end of 1941. In the foreground, the administration building is in the center, flanked by Hangar Row, with Dead Horse Bay in the foreground. Beyond the runways, the Coast Guard air station and the Neutrality Patrol seaplane base can be seen on the Jamaica Bay shoreline. (NA.)

THE NAVY TAKES OVER.
Shortly after Pearl Harbor,
the Navy purchased
FBF outright from New
York City for $9.25
million. Even prior to the
purchase, the air station
had already embarked
on its main task: ferrying
new aircraft from local
factories to combat units,
as well as for shipment
to England. The lineup
of Grumman, Vought,
and Douglas aircraft
testifies to the magnitude
of that task. (NPS.)

NAVY MAKES CHANGES TO FBF. After the Navy took over, the remaining residents were evicted from the Barren Island community, shown in the foreground. Eventually, all the structures were demolished and replaced with barracks and other Navy buildings. The bridge in the upper left is the Marine Parkway Bridge, recently renamed for Brooklyn Dodger first baseman Gil Hodges. (NA.)

NAVAL AIR STATION NEW YORK (NASNY). By late 1942, the original runway 6/24 was closed since its threshold was dangerously close to structures. A new runway was built on the north side of the field, and the others were widened and lengthened. The administration building and Hangar Row line Flatbush Avenue, which runs across the photograph above. In the August 1943 view below, the continued development of NASNY is obvious, including many Navy structures in the former Barren Island community. The far end of the bridge is the site of the former Rockaway Beach Reserve base. In the foreground, lining the shore from left to right are the CG Air Station, Hangar A, and Hangar B. (Above, NNAM; below, NA.)

Navy Winter Garb. In this February 1942 photograph, five fliers and crewmen model cold-weather flight suits and uniforms in front of a Navy J2F Duck. Both Navy and Coast Guard pilots flew these Grumman amphibians—with a crew of pilot, observer, and radio operator—on antisubmarine patrols and rescue missions from FBF. (NA.)

PBY at NASNY, 1942. The Consolidated PBY Catalina flying boats and amphibians were perhaps the most important patrol aircraft of the war. More than 3,300 were built in the United States and under license in Canada. Navy PBYs flying from FBF flew endless antisubmarine patrols during the war, accounting for a number of probable U-boat kills. They also rescued countless survivors of ships sunk by the German submarines. (NA.)

PV-1 Ventura, February 1943. In addition to the seaplanes and amphibians, the Navy also flew land-based antisubmarine patrols from FBF using Lockheed PV-1 Venturas, like the one shown here, and later PV-2 Harpoons. Lockheed supplied a long line of land-based antisubmarine aircraft to the Navy, a monopoly that was recently broken by Boeing's P-8A Poseidon. (NA.)

Coronado at Dusk, FBF 1944. While the PBY Catalina dominated Navy contracts for seaplane and amphibian patrol planes, Consolidated's PB2Y Coronado flying boat offered much longer range; however, with a price triple that of the PBY, the Navy bought limited numbers and used them mainly for transoceanic transport. (NA.)

NAVAL AIR FERRY COMMAND, NASNY. As production of warplanes increased, so did the need to ferry them from factories to combat squadrons. Since FBF was surrounded by important factories producing Navy aircraft, it was a logical focal point. New aircraft were flown by Navy crews to FBF, where they were tested and outfitted with equipment and then ferried to the fleet, mainly at West Coast bases. The insignia (below) of the Navy's FBF-based Air Ferry Service Squadron One (VRF-1) shows a stork delivering a "newborn" Grumman Hellcat westward. The route involved numerous stops in the Southeast and Southwest; an Air Ferry Service pilot plots his route (at left) to the West Coast prior to departure. During the war, 75,000 aircraft were ferried by the FBF-headquartered Naval Air Ferry Command, 46,000 through FBF alone. (Both, NNAM.)

New Corsairs Ready for Delivery. Virtually all Navy aircraft built in Northeastern factories, including Grumman, Eastern, Chance Vought, Vought Sikorsky, Brewster, and Martin, were delivered through NASNY at FBF. These F4U-1A Corsairs were produced in Vought's Stratford, Connecticut, plant 50 miles to the northeast. The Corsair was the longest-produced fighter of World War II and achieved an 11-to-1 kill ratio. (NPS.)

Curtiss SB2Cs of VRF-1. The last in a long line of Curtiss Helldiver–named aircraft was the SB2C-5, like the ones shown here returning to FBF with air ferry crews. The designation indicates the second version of the SB model scout bomber built by Curtiss, but Navy aircrew referred to it as the "son-of-a-bi*ch-second class" because of its difficult flight characteristics. (TH.)

VOUGHT PRODUCTION OF CORSAIRS. The F4U Corsair was built in Chance Vought's Stratford, Connecticut, factory (above), a short hop from FBF. Located on the north shore of the Long Island Sound, the site is now Sikorsky Memorial Airport. Vought moved out of its original Long Island City factory in 1939. A total of 7,829 Corsairs were built at Stratford between 1940 and 1952. Female assembly-line workers (below) are shown testing the functioning of the landing gear of a F4U-1A Corsair; women played a vital role in wartime production of aircraft, ships, and equipment, taking the place of men serving in the military. (Above, NA; below, Vought.)

"GRUMMAN IRON WORKS." Grumman's factory in Bethpage, Long Island, seen above, was called the Iron Works because of the robustness of its products. It was just 20 miles east of FBF. That factory produced an astounding 12,275 Hellcats between 1942 and 1945. Hellcats achieved a 19-to-1 kill ratio and destroyed more enemy aircraft than any other Allied fighter. Air Ferry Service pilots from FBF flew new warplanes from factories to the fleet and also returned "war-weary" aircraft for refurbishment. The Hellcats pictured below are being rebuilt by Grumman after their return from combat; they will be put back in service, usually in the training of new pilots. (Above, NA; below, NNAM.)

EASTERN AIRCRAFT, LINDEN. When Grumman introduced the F6F Hellcat in 1942, demand was so great that it was forced to subcontract the continued production of its earlier F4F Wildcat fighter and TBF torpedo bomber. Grumman subcontracted with Eastern Aircraft, a subsidiary of General Motors, to produce 1,060 FM-1s and 4,777 FM-2s (an improved version of the Wildcat). The GM plant in Linden, New Jersey, seen above, formerly produced Buick, Pontiac, and Oldsmobile automobiles. Its location, 20 miles west of FBF, was ideal since it was adjacent to Linden Airport. Below, newly produced FM-2 Wildcats are shown lining the Linden ramp. (Above, NA; below, NNAM.)

EASTERN AIRCRAFT, TRENTON. At the same time that Wildcat production started in Linden, Grumman subcontracted production of the Avenger to Eastern Aircraft as well. GM converted a parts factory in Trenton, New Jersey, 50 miles southwest of FBF, for the production of its version of the TBF Avenger torpedo plane, known as the TBM. The factory (above) produced 7,540 TBMs by the end of the war. It was selected because of its co-location with Trenton Airport, also known as Naval Air Facility Mercer. The lineup of produced TBMs sitting outside of the factory (below) gives an indication of the facility's productivity. (Above, NA; below NNAM.).

NAS New York, April 1944. This aerial view shows the expansion of the naval air station, including hangars added on to Hangar Row (lower right) to house the Air Ferry Service. Note the large number of aircraft parked on the ramp outside the hangars. The seaplane and Coast Guard bases are on the Jamaica Bay (left) side. (NA.)

Grumman Goose of VRF-1. Air Ferry Service squadrons relied on utility aircraft, such as this Grumman JRF Goose, to shuttle pilots to the various factories to pick up new aircraft to ferry to FBF for testing and outfitting of special equipment. The JRF Goose had a crew of two and could carry up to seven passengers. (TH.)

NASNY'S AUXILIARY AIRFIELDS. With so much air traffic into and out of FBF, NASNY also operated two naval auxiliary airfields (NAFs). NAF Roosevelt Field (above) in Mineola, Long Island, was historically significant as point of origin for the flights of Charles Lindbergh and other notable aviators. It was used for the servicing and air testing of new aircraft destined for the Royal Navy. NAF Mercer Field (below) in Trenton, New Jersey, was eventually used for the acceptance, equipping, and testing of FM Wildcats and TBM Avengers produced by Eastern Aircraft. Today, there is no obvious evidence of the historical importance of Roosevelt Field—it is a shopping center. Mercer Field is now Mercer County Airport, mainly a private and corporate aircraft base with, at the time of this writing, airline service by Frontier Airlines. (Both, NA.)

NATS' DOUGLAS SKYMASTER. From March 1942, FBF became the East Coast terminus for the Naval Air Transport Service (NATS). As the war progressed, so did NATS operations, with two squadrons based at NASNY providing both transcontinental and transatlantic passenger and cargo service. In a typical month, 700 flights would carry more than 5,000 passengers and 5,000 tons of cargo. One of the mainstays of NATS was the Douglas R5D Skymaster, the military version of the DC-4 airliner. More than 200 R5Ds went into service with the Navy and Marines during the war, with many staying in service through the 1960s. Here, a Skymaster is shown in flight (above) and being loaded with oversized cargo at FBF (below). (Both, NA.)

PAN AMERICAN CLIPPER. In late 1943, a naval overseas air cargo terminal was built on the east side of FBF. Run under contract by Pan American World Airways' Atlantic Division, Boeing 314 Clippers (shown here) and Consolidated PB2Y Coronado flying boats hauled all transatlantic cargo traffic from that base. (TH.)

CHANGE OF COMMAND CEREMONY. Capt. Kenneth Whiting was a naval-aviation pioneer sometimes referred to as the "father of the aircraft carrier." On February 25, 1943, he relieved Capt. Ernest McDonnell as commander of NAS New York. Unfortunately, Whiting suffered a fatal heart attack exactly two months later. Whiting Field, Florida, where today's naval aviators begin their flight training, is named in his honor. (NA.)

WAVES IN REVIEW, JULY 1944. In 1943, NAS New York's complement was bolstered by the influx of 42 officer and 350 enlisted WAVES (Women Accepted for Voluntary Emergency Services). Although the all-woman division of the Navy was established in July 1942, it took until June 1948 for women to gain permanent status in the US military. In the photograph above, WAVES commanding officer Lt. Comdr. Mildred McAfee (left) and NASNY commander Capt. Newton White review the WAVES on the first anniversary of their establishment. Rows of new Grumman Hellcats, with the temporary manufacturer's numbers on their noses, are in the background of the photograph below. (Both, NA.)

WAVES AT NAS NEW YORK.
Above, a contingent of WAVES
based on FBF is shown marching
past Hangars 3 and 4, with the sick
bay on the right. The administration
building, which became the
Naval Air Station headquarters
building, is just out of view to the
right. The facade of the hangars
underwent many changes over
the years, as the photographs in
this book can attest. In wartime
service, the WAVES performed a
range of duties, including air-traffic
control and control tower duty (at
right), administrative jobs, and
responsibilities as aircraft mechanics,
crew chiefs, parachute riggers, and
more. (Above, NA; at right, NPS.)

"Miss Air Wave 1944." The word "emergency" in the full title of the WAVE division indicated that women's acceptance in the military was for the circumstances of the war, not a permanent position; regardless, their contribution was immeasurable. Frances Doyle, USN storekeeper second class, shown posing on a North America SNJ, may have won the base beauty contest, but her real job was in the supply department. (NA.)

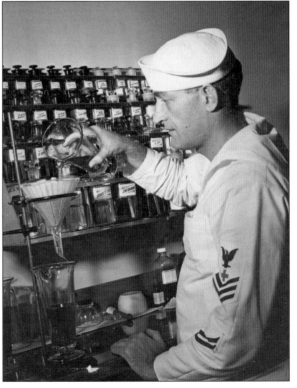

Navy Sick Bay, November 1942. Naval air stations traditionally are considered the equivalent of ships, hence the base's medical department and pharmacy was located in the station's sick bay. At FBF, it was located in a structure built between the administration building and Hangars 3 and 4. Here, USN pharmacist Mate First Class Henry E. Randall is filling a prescription for a base sailor. (NA.)

NAVY BAKERS, NOVEMBER 1942. Barracks and mess halls were constructed at the southern end of FBF, where the Barren Island community had been located prior to the Navy takeover of the airfield. In this photograph, Navy bakers aboard NASNY are preparing for the next meal. As the station's complement swelled, housing was inadequate, and many sailors had to live in apartments off base and commute via public transport. (NA.)

COAST GUARD AIR STATION, 1944. As the size of NASNY expanded, so did Coast Guard Air Station Brooklyn. In addition to the Coast Guard's traditional role of search and rescue, the primary wartime mission became antisubmarine patrols. This photograph shows new structures added since commissioning, as well as a PBY Catalina anchored offshore and Curtiss SO3C Seamew floatplanes taxiing down the ramp past a Hall PH-3 seaplane. (NNAM.)

INTRODUCING THE HELICOPTER, JULY 1943. While the USCG's antisubmarine and life-saving activities were vital, CG Air Station Brooklyn's most historic achievement was the introduction into service of the newly invented helicopter. In this photograph, from left to right, Comdr. George H. Bowerman, Coast Guard commandant Adm. Russell R. Waesche, and Lt. Art Hesford are about to witness a helicopter demonstration at FBF. (NNAM.)

HELICOPTER DEMONSTRATION, CGAS BROOKLYN. On July 10, 1943, Coast Guard and Navy personnel at FBF got their first glimpse of a helicopter. A Sikorsky YR-4B helicopter, on loan from the Army Air Corps, put on a demonstration in front of the Coast Guard hangar while conventional aircraft line the ramp. The body language of the observers gives an indication of how exciting that event must have been. (NNAM.)

Igor Sikorsky. Russian émigré Igor Sikorsky founded his airplane company in 1923. Originally focusing on flying boats, including the first used in commercial flights by Pan American World Airways, he built the first viable American helicopter prototype in 1939. In 1942, this became the world's first mass-produced helicopter, the R-4. He is shown here being greeted by Coast Guardsmen at Air Station Brooklyn in July 1943. (NNAM.)

USCG Commander Frank Erickson. In May 1942, then Lieutenant Erickson was transferred to CGAS Brooklyn and immediately ordered to Sikorsky Aircraft's plant in Bridgeport, Connecticut, for training in the construction and operation of helicopters. He became the Coast Guard's first helicopter pilot and its leading advocate. Back at FBF, he commanded the base that trained 102 pilots and 225 mechanics for the US and British militaries. (NA.)

USCG LIEUTENANT STEWART GRAHAM. Lieutenant Graham was Commander Erickson's protégé and the Coast Guard's second helicopter pilot. Like his mentor, Graham learned to fly helicopters on Sikorsky YR-4Bs on loan from the Army before the Coast Guard received its own version, the HNS-1 Hoverfly. Training new pilots during hot summer days proved problematic as these early helicopters lacked sufficient power to lift two pilots under such conditions. (NA.)

LIKE FATHER, LIKE SON. Igor Sikorsky's 19-year-old son Sergei followed in his father's footsteps, enlisting in the Coast Guard in 1943 for the express purpose of working on helicopters produced for the military by his father's company. In this March 1944 photograph, Coast Guard Machinist Mate Sikorsky is shown waxing a rotor blade of a HNS-1 helicopter. (NA.)

HELICOPTER DEMONSTRATIONS AT FBF, 1944. Commander Erickson and Lieutenant Graham established the versatility of helicopters in addition to training new pilots and ground crew. Search and rescue, the traditional Coast Guard task, was paramount. In the image at right, Erickson is shown hovering a HNS-1 while crewmen transfer a survival radio into the craft, demonstrating the stability of the helicopter while equipment is loaded or off-loaded at a simulated rescue site. Since autopilots for helicopters had not yet been invented, flying the already-precarious machines in bad weather or at night was quite challenging. In the time-exposure photograph below, night takeoffs and landings as part of new pilot training are being demonstrated. (Both, NA.)

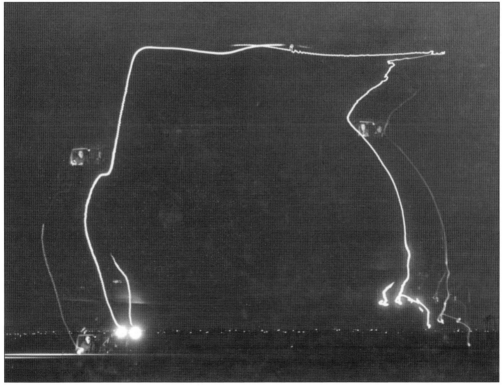

RESCUE HOISTS—ANOTHER INNOVATION. Recognizing the life-saving potential of helicopters, Commander Erickson, Lieutenant Graham, and Coast Guardsman Sergei Sikorsky teamed up with Igor Sikorsky to develop and demonstrate a powered hoist system. At left, Erickson is piloting a HNS-1 Hoverfly, while Igor Sikorsky is strapped to the hoist, about to be lowered to the ground at FBF on August 14, 1943. Shortly thereafter, Erickson expanded his rescue demonstrations to include picking up "survivors" from the water or from a speeding boat. The photograph below shows such a pickup from Jamaica Bay; Hangars A and B of the seaplane base are in the background. (Both, NNAM.)

HELICOPTER TRAINING, 1944. Once training got under way, the Coast Guard air station's HNS-1 Hoverfly helicopters flew daily at FBF and surrounding locations. Above, British army major John Richardson (left) is receiving instruction from Coast Guard lieutenant Alan Kleisch at Rockaway Beach, with another helicopter hovering behind. Located across Jamaica Inlet from FBF, the former Naval air station allowed helicopter-landing practice away from the heavy air traffic of the main air station at FBF. Below, another HNS-1 is returning to the Coast Guard air station. The complex rotor mechanism invented by Sikorsky and shown in the foreground allows for changes in the pitch of the rotor blades for lift and angle of the mast to control direction of flight. (Both, NA.)

MORE HELICOPTER OPERATIONS. The Coast Guard air station was tasked to train helicopter ground crew as well as aircrew. At left, Coast Guardsmen are refueling a HNS-1 Hoverfly from an underground fuel hydrant. Below, ground crewmen await the return of one of their "birds." First equipped with three YR-4B prototypes obtained from the Army, the Navy accepted the first of 23 HNS-1s on behalf of the Coast Guard in October 1943. Power was supplied by a meager 200-horsepower Warner R-550-3 engine, giving the Hoverfly a top speed of only 82 miles per hour. (Both, NA.)

ANTI-TORQUE ROTOR. The Sikorsky HNS-1 was equipped with a 38-foot-diameter, two-bladed main rotor. A small rotor on the tail, another Sikorsky innovation, was used to counteract the torque generated by the main rotor. Here, ground crewmen are examining the antitorque rotor while another HNS-1 hovers nearby. (NA.)

HELICOPTER SIMULATOR. Compared to today's full-motion visual display simulators that are widely used in flight training, the helicopter simulator installed at Coast Guard Air Station Brooklyn and shown in this photograph seems archaic and almost comical. A full-sized HNS-1 cockpit was suspended by a trapeze in an attempt to give student pilots a feel for helicopter flight and its controls. (USCG.)

SIMULATING SHIPBOARD OPERATIONS. In an effort to expand the utility of helicopters, Commander Erickson sought to demonstrate shipboard operations. In order to train helicopter pilots in the risky takeoffs and landings from a ship at sea, a moving platform simulating a ship's deck was built at FBF. The platform, seen above, was humorously named USS *Mal De Mer* because of its rocking motion. Below, a float-equipped HNS-1 is practicing landing on the platform. Visible beneath the nose of the helicopter is the administration building. To the left are the paired Hangars 3/4, and to the right are Hangars 5/6 and 7/8. Still farther right is Hangar 9, built for use by the Naval Air Ferry Command. (Both, NNAM.)

USS TURNER TRAGEDY.
The helicopter gave its first real-life demonstration of its life-saving potential on a cold, snowy January 3, 1944. On that day, the USS *Turner* (above), anchored off New York Harbor, suffered a series of internal explosions that caused the two-year-old destroyer to capsize and sink, taking 15 officers and 123 crew members with it. Survivors were picked up by nearby ships, and a Coast Guard Sikorsky HNS-1 piloted by Comdr. Frank Erickson flew two cases of blood plasma lashed to the helicopter's floats from FBF to Sandy Hook, thus saving the lives of many of *Turner's* survivors (at right). (Both, NA.)

INSPECTION AT AIR STATION BROOKLYN. Personnel and six HNS-1 Hoverfly's are lined up for inspection outside Coast Guard Air Station Brooklyn in this late-1944 photograph. Interestingly, the ramp is devoid of conventional aircraft, although they still flew from the base. The structures in the background include the large Hangar A (built originally for the Navy patrol plane base), the station's powerhouse, and numerous storage buildings. (NNAM.)

HOVERFLY OVER LOWER MANHATTAN. The Manhattan skyline is clearly visible from FBF, and this proximity makes it a tempting background for air-to-air photographs of aircraft stationed there. In this image, a Coast Guard HNS-1 is posing over the lower East River near Wall Street and the present-day South Street Seaport. (NA.)

DAY-TO-DAY OPERATIONS. Coast Guard Air Station Brooklyn was an independently operating entity located within NAS New York, both situated at Floyd Bennett Field. It provided for daily operations, including scheduling of aircrew and maintenance, separate from their Navy neighbors. At right, Coast Guard lieutenant commander Red Bolton, duty officer at the Air Station, checks the status board of station officers. Note the use of the nautical term "aboard" to indicate the officer is present at the air station. The image below shows the instrument shop, where aircraft instruments such as Sperry gyroscopes were maintained and repaired. (Both, NNAM.)

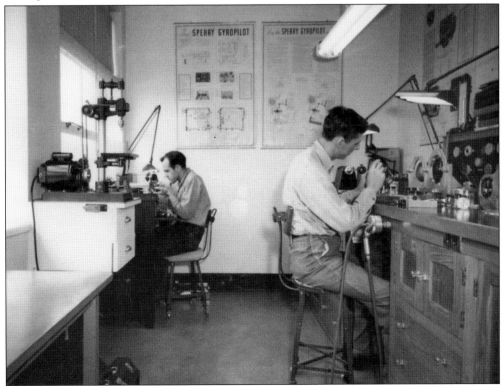

Hall PH-3s in Wartime Markings. With the outbreak of hostilities, aircraft in all the services lost their colorful markings as they were replaced with drab camouflage colors. For the Coast Guard, prewar bright aluminum finishes with red, white, and blue rudder colors gave way to nonspecular blue-gray on surfaces viewed from above, and nonspecular light gray on those seen from below, as shown in these photographs of Coast Guard PH-3s at Air Station Brooklyn. The image above shows one afloat at the foot of the seaplane ramp in 1944, with the hangar in the background. Note that the insignia includes a star surrounded in blue, with white rectangles. Below, depth charges are being loaded for an antisubmarine patrol in 1943, showing the earlier star insignia minus the rectangles. (Above, NNAM; below TH.)

COAST GUARD AIRCREW AT STATIONS. The mainstay of Air Station Brooklyn in the early war years was the Hall PH-3 flying boat. Powered by two 750-horsepower Wright R-1820F-51s, its top speed was 159 miles per hour with a range of 1,937 miles. It was armed with four flex-mounted .30-caliber Lewis guns and could carry 1,000 pounds of bombs. The crew of six included a pilot, copilot, navigator, radioman, and two gunners/observers. In the photograph at right, a crewman demonstrates aerial photography from one of the aircraft windows, while the image below shows a gunner aiming through the sight of his Lewis machine gun. (Both, NNAM.)

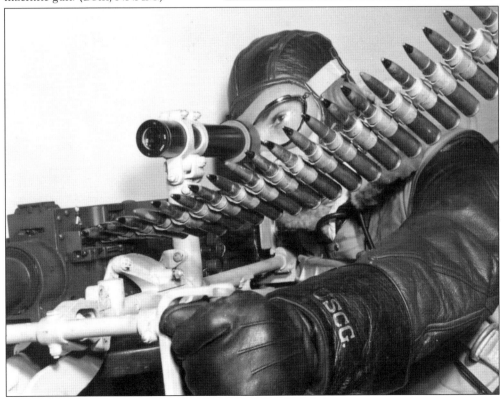

CURTISS SOC-4. This Coast Guard Curtiss SOC-4 Seagull floatplane is pictured on its land dolly at Air Station Brooklyn, FBF, in 1943. Although other station aircraft had been repainted with wartime gray colors, this example still wears its bright prewar colors, including the unique red, white, and blue–painted rudder. (NNAM.)

GRUMMAN J2F-2. This Grumman J2F-2 Duck also has yet to be repainted in drab wartime colors. Note how the Duck amphibian's fuselage includes a hull for landing on the sea; therefore, it is an amphibian seaplane. In contrast, the SOC-4 Seagull shown in the previous photograph has a separate fuselage and float, hence it is a floatplane. (NNAM.)

GRUMMAN J4F-1, 1943. The Coast Guard purchased 25 Grumman J4F-1 Widgeon amphibians to replace the early biplanes, with delivery starting in July 1941. It was basically a civilian design adopted for military service with the addition of a bomb rack on each wing. While Grumman fighters were named after cats, their amphibians were named after seabirds. (NNAM.)

CURTISS SO3C-3. This photograph, taken in 1943 outside of Air Station Brooklyn's hangar, shows a Curtiss SO3C-3 with the optional wheeled landing gear. The Seamew and other floatplanes of the time were designed with interchangeable floats and wheels. The Seamew replaced the earlier Curtiss Seagull in service. (NNAM.)

USS *LAFAYETTE*, FEBRUARY 1942. The French luxury liner *Normandie* sought haven in New York harbor at the outbreak of World War II. The US government confiscated the ship and renamed it the USS *Lafayette,* intending to convert it to a troopship. While it was docked at New York City's Pier 88 for the conversion work, sparks from a cutting torch ignited a pile of highly flammable life preservers, starting a fire that would gut the ship. Firefighters poured 6,000 tons of water into the top decks, causing the ship to eventually capsize. The photograph above shows a Coast Guard SO3C from Air Station Brooklyn flying over the hulk after the superstructure has been removed. In the image at left, Mayor Fiorello LaGuardia confers with fire marshal Patrick J. Walsh at the disaster site. (Above, TH; at left, NA.)

Four

NAS NEW YORK
IN THE COLD WAR

HIGH SCHOOL STUDENTS VISIT. The events of the World War II stimulated interest in aviation among many young people. Here, a group from New York City's Stuyvesant High School tours the naval air station and closely examines a Consolidated PBY-6A Catalina amphibian patrol plane of the Naval Air Reserve Unit at FBF in October 1952. (NA.)

"Flying Pancake" at FBF. The years after World War II saw a drastic drop in aircraft factory output, and the Air Ferry Service was disbanded; however, a great number of interesting types were evaluated during that period, including this experimental Vought V-173 "all wing" design. Promising initial results led to a larger version, the XF5U "Flying Flapjack," which was built but never flown before being cancelled. (Vought.)

July 4, 1948. The 1928 committee selecting the municipal airport site preferred a tract on the eastern end of Jamaica Bay to Barren Island. In 1946, that site was chosen for Idlewild Airport, today's JFK. Hundreds of aircraft gathered at FBF for the opening-ceremony air show, including (from the front) a North American FJ-1 Fury, the prototype Grumman XF9F-2 Panther, and two McDonnell FH-1 Phantoms. (TH.)

GRUMMAN HELLCATS OF NARU-NY. In the postwar and Cold War years, NAS New York was the largest NARU base in the country, supporting at its peak 3,000 Navy and Marine reservists. At the start, the Reserve fighter squadrons were equipped with F6F-5 Hellcats. The initial markings for all Reserve aircraft included a glossy international orange band around the fuselage and a large aircraft number (above). In 1947, more markings were added, including the stars-and-bars insignia, and the "R" tail code for New York, as shown on this F6F-5P (below). Prior to that change, the "RF" tail code was temporarily used. (Both, TH.)

VOUGHT CORSAIRS OF **NARU-NY.** Flying alongside the Hellcats were old Vought F4U-1D and license-built Goodyear FG-1G Corsairs. Above, a Corsair in original NARU markings is taxiing at FBF in front of a North American SNJ station "hack" used as a utility aircraft. The image below shows an example of "ramp rash," the consequence of a relatively minor ground collision on the ramp. Both Corsairs in this image wear the later Air Reserve markings with the "R" tail code designating New York. By 1949, all Navy Air Reserve Unit fighting squadrons were reequipped with newer model Vought F4U-4 Corsairs. (Above, TH; below, NA.)

NARU-NY PHANTOMS. Although McDonnell Aircraft built only 62 FH-1 Phantoms (not to be confused with that company's later F4H Phantom II, of which more than 5,200 were built), they did equip the Navy's first operational carrier-based jet squadron. After a short time, they were replaced with newer jets, and the FH-1s reverted to Naval Air Reserve squadrons, such as Fighting Squadron 842 (VF-842) at NAS New York. The September 1950 photograph above shows officers and enlisted personnel in front of one of their Phantoms, while the April 1951 photograph below shows a group of Reserve FH-1s overflying their favorite backdrop: Manhattan. (Above, TH; below, NA.)

KOREAN WAR ACTIVATION. As a result of the outbreak of the Korean War, three Navy and two Marine Air Reserve squadrons based at NAS New York were activated in February 1951. All personnel of Navy Fighting Squadrons 831 and 837 (VF-831 and VF-837) were called to active duty, and the squadrons flew combat missions on USS *Antietam's* September 1951 to March 1952 war cruise. In the photograph above, VF-837/831 mechanics are checking a Pratt & Whitney J-42 engine of a F9F-2 Panther on the hangar deck prior to a mission. Below, the deck edge elevator is bringing a VF-831 Panther up to the flight deck from the hangar deck. (Both, NA.)

PREPARING PANTHERS FOR A MISSION. The two NAS New York Panther squadrons pooled their aircraft for maintenance and missions, but the squadrons' aircraft could be differentiated by their modex, or visual identification system code numbers. VF-831 aircraft were numbered in the 300 series, while VF-837 aircraft were in the 400 series. In the photograph above, a deck crew is maneuvering a VF-837 Panther on the flight deck using a "mule," Navy parlance for a tractor. Below, red jersey–clad ordinance men are loading Panthers (left) and AD Skyraiders (right) with bombs and rockets prior to a mission over Korea. (Both, NA.)

PREPARING FOR A MISSION. Prior to a mission, Brooklyn's VF-831 and VF-837 Panthers are arranged on *Antietam's* flight deck (above). Normally, piston-engined F4U-4 Corsairs and AD-4 Skyraiders would be launched first, while the faster jets would be catapulted off the carrier's deck last but in time for the strike group to rendezvous over the target. While the aircraft are being prepared, pilots attend a briefing in their squadron's ready room (below). Shown from left to right are (first row) Lt. Comdr. William Ryan and Comdr. Anthony Denman; (second row) Lt. William Betz and Lt. George Asip. (Both, NA.)

MORE MISSION PREPARATION. In the February 1952 image seen above, Brooklyn-based reservists Lt. Julian Forster (left) and Lt. Lemeul Arnold plan routes to a target near Hamhung, North Korea, from *Antietam*'s location in the Sea of Japan. At right, with the briefing complete, pilots scramble up ladders from the ready room, located off the hangar deck, to their waiting aircraft on the flight deck. Wearing rubberized cold-weather emersion suits, the pilots worked up quite a sweat just getting to their aircraft. Later carriers were equipped with escalators to provide a bit more comfort to the pilots manning their aircraft. (Both, NA.)

MOUNT UP AND LAUNCH! Reaching the flight deck, pilots find their assigned aircraft based on the large modex numbers, and with the help of plane captains wearing brown jerseys, they climb into their cockpits and strap in (above). The carrier is picking up speed, as can be seen from the black smoke emanating from the stack, and will turn into the wind for launching aircraft. While propeller planes, like the Corsair and Skyraider, simply flew off the decks, heavier jets required the use of one of the two catapults (below). On *Antietam* and other World War II carriers pressed into service for the Korean War, the old hydraulic catapults were barely adequate to launch a fully laden Panther and were the cause of many accidents. (Both, NA.)

OFF TO THE TARGET, DECEMBER 1951. The goal is to launch all the mission jets as quickly as possible so that the first off do not have to waste fuel circling as they wait for the subsequent jets. On a good day, the two hydraulic catapults could launch two aircraft per minute (above). Despite the existence of the hydraulic catapults, the carrier still needed to steam at top speed into the wind to generate the highest possible "wind over the deck" for launches. On the way to the target, F9F-2 Panthers of VF-831 and VF-837 fly in formation with a F9F-2P photoreconnaissance Panther of Composite Squadron 61 (VC-61), also from *Antietam* (below). (Both, NA.)

COMPLETING THE MISSION. Returning aircraft are guided to their landing by hand signals and verbal cues given by the landing signal officer (LSO) stationed at the aft end of the flight deck (above). The *Antietam*, typical of the straight deck carriers of that era, was equipped with 12 cross-deck arresting cables known as "wires." The goal was for the tailhook of landing aircraft to catch one of these cables, which would then play out, bringing the aircraft to a halt. A "deck runner" wearing a yellow jersey would then sprint across the deck to assist in disengaging the hook from the cable (below), allowing the jet to fold its wings and taxi to the forward parking area before the next jet landed. (Both, NA.)

OUCH! Flying jet aircraft from World War II–vintage aircraft carriers was a daunting task; the accident rates made operations around the carrier more perilous than actual combat with the enemy. If a landing aircraft failed to hook one of the cross-deck cables, two barriers were designed to catch it before it plunged into the aircraft parked at the front end of the deck. Above, a VF-837 Panther is shown after such a barrier engagement, which caused the nose gear to collapse. Airframe mechanics had to work feverishly overnight to repair aircraft from things such as flak damage, seen below, for the next day's mission. (Both, NA.)

MISSION DEBRIEF ABOARD ANTIETAM, FEBRUARY 1952. For the pilots, a mission is not complete until their debriefing with the squadron commander and intelligence officers. These Brooklyn reservists go into a debriefing session and check themselves on targets hit and damaged. Above, from left to right, Ens. Robert Thomas, LTJG Robert Baker, and Ens. Howard Moehn compare notes. Below, in the squadron ready room are, from left to right, Lt. Comdr. Wilbert Hackbarth, LTJG Joseph Voda, Lt. George Johnson, and Lt. Gavin Weir. (Both, NA.)

BACK AT FLOYD BENNETT FIELD, DECEMBER 1953. In addition to the five Reserve squadrons activated for the Korean War, the remaining squadrons at NASNY continued with their training and other assigned tasks. One task was to greet returning Korean War wounded and transport them to local hospitals, including the St. Alban's Veterans Administration Medical Center in Queens, New York. In the image above, Naval Air Transport Service (NATS) Douglas R5D Skymasters line the ramp of NASNY, while in the image below, returning wounded Korean War veterans are assisted out of a Skymaster to waiting ambulances. (Both, NA.)

"PREGNANT GUPPY" AVENGERS. The Grumman/Eastern Avenger was one of the best carrier-based bombers of World War II. After the war, some survivors were converted to specialized tasks, such as these TBM-3W early warning aircraft of Antisubmarine Squadron 833 (VS-833) at NASNY in 1954. The nickname "pregnant guppy" came from the obvious bulge housing an ANS-20 radar antenna. (TH.)

NAS NEW YORK, SUMMER 1953. At its peak, 34 Navy, Marine, and joint-service squadrons called NASNY their home. These included fighting, attack, antisubmarine warfare, patrol, transport, and utility squadrons. In this view looking northwest from Jamaica Bay, the Coast Guard air station is in the foreground. To the right, the two large hangars, A and B, house the many Reserve squadrons. (NPS.)

Aerial Perspectives of Floyd Bennett Field. Comparison of the image at right, taken in 1957, with earlier photographs in this book shows the extent to which NASNY expanded. The three widened and lengthened runways are apparent, as is the complex of barracks, mess halls, repair shops, and more in the right foreground. The Coast Guard air station and Navy/Marine Reserve headquarters are along the shore to the right. The photograph below shows the original administration building and Hangar Row looking east; the dark gray building to the left of the administration building is a "nose hangar" built for the resident New York Air National Guard's 106th Wing. (At right, NA; below, NPS.)

NAS NEW YORK HELICOPTERS. After the initial and critical development work performed by Coast Guard Air Station Brooklyn during World War II, there was a short lapse of military interest in helicopters until their utility was demonstrated during the Korean War. Reserve units at NASNY included helicopter squadrons for search and rescue, antisubmarine warfare, and utility work. The Piasecki (later Vertol) HUP-2 Retriever in the photograph above is taking off on a rescue mission to Martha's Vineyard in September 1954. Below, a Sikorsky HRS-1 utility helicopter is showing off its front-mounted Pratt & Whitney R-1340 engine during an open house. (Above, NA; below, TH.)

NAS NEW YORK PATROL AIRCRAFT. Naval Air Reserve patrol squadrons at Floyd Bennett Field flew two types of aircraft during the Cold War years. The Lockheed P2V Neptune (above) was the mainstay of the Navy's land-based patrol squadrons. A number of Reserve squadrons at FBF shared a pool of eight Neptune patrol planes. Patrol Squadron 832 (VP-832) was activated in October 1961 during the Berlin Crisis, deploying with four P2V-5Fs to NAS Guantanamo Bay, Cuba. Another Reserve squadron, VP-831, was called up for the Cuban Missile Crisis in October 1962, deploying with Neptunes to NAS Patuxent River. The Grumman S2F-1 Trackers shown over the very photogenic Manhattan skyscrapers (below) in April 1957 equipped several antisubmarine Reserve squadrons at FBF, including VS-833. Note the newer "7R" tail code for New York reserves. (Both, TH.)

NAVY RESERVE GRUMMAN COUGARS, MAY 1954. In the mid-1950s, the NARUs became part of Naval Air Reserve Training Command (NART). At that time, the inventory of NART aircraft based on NAS New York numbered 90. The predominant aircraft was the Grumman F9F-7 Cougar, the swept-wing successor to the Panther. In the image above, Aviation Augmentation Unit 836 (AAU-836) pilots—from left to right, Lt. S. Coddington, Lt. F.L. Zito, Lt. J.E.H. Schreiber, and Lt. J. Organek—look over a map with their route for a cross-country training flight. At left, Lt. S. Coddington (left) and Lt. F.L. Zito walk down the ramp to their waiting Cougars. (Both, NA.)

GRUMMAN COUGARS OF NART NEW YORK. Once again, the alluring Manhattan skyline is used for exciting aerial photography. The March 1955 photograph above shows four FBF-based Reserve fighting squadron F9F-7 Cougars flying past the spires of the New York Life (left) and the Metropolitan Life (right) insurance company headquarter buildings. In the June 1954 photograph below, a dark sea-blue NART F9F-7 Cougar forms the backdrop for the unit's maintenance officers, flanking Comdr. H.C. Jipson (second row, sixth from the left) and Capt. J.S. Anderson (second row, seventh from the left). Grumman built a total of 1,392 Cougars in its nearby Bethpage plant. First flown in 1951, the Cougar remained in service until 1974, when the last retired from the Navy's Training Command. (Both, NA.)

CHANGE OF PAINT SCHEME THEN AIRCRAFT. One interesting aspect of naval aviation is the constantly changing spectrum of aircraft color schemes. In 1955, a brighter gull gray over white paint scheme, as shown above of a NART-New York Cougar, replaced the overall dark sea-blue scheme. It took a while for all aircraft to be repainted, so there were many occasions of squadrons flying a mix of aircraft colors. By 1961, the Cougars were replaced by North American FJ-3 and then FJ-4B Furies, as shown below. Three Navy and Marine Reserve attack squadrons shared a common pool of Furies, as indicated by the joint "Navy-Marine" markings. The international orange trim designated Reserve squadron aircraft. Douglas A4B and A4C Skyhawks, shown in the left background, replaced the Furies. (Both, TH.)

LOCKHEED SEASTAR, NART-NEW YORK. Three Lockheed T-1A Seastars were assigned to the joint Navy-Marine Reserve units during the 1960s and were used for proficiency training and as general "hack" aircraft. The T-1A, earlier known as the T2V, was a radically modified and "navalized" version of the Air Force T-33 trainer. (NNAM.)

OPEN HOUSE, JULY 1954. The Navy always made an effort to maintain a good relationship with its Brooklyn neighbors. One opportunity was the *New York Daily Mirror*'s Annual Model Aircraft Meet. Here, a Naval Reserve K-series patrol blimp from NAS Lakehurst overflies the event that coincidentally was attended by this book's author at age nine. (NA.)

ANNUAL INSPECTION, 1954. Once a year, all air station personnel were gathered for the annual inspection. In the November 1954 inspection, R. Adm. Daniel Gallery presided. NAS New York boasted two enormous hangars built on the east side of the field, each of which could hold the assembled officers and men of the air station, along with their guests. (NA.)

FBF RECORD-BREAKING FLIGHTS RESUME. The central role of Floyd Bennett Field in the many record-breaking flights of the 1930s resumed after the war. In 1946, Lockheed P2V-1 Neptune *Truculent Turtle* broke the nonstop distance record flying from Perth, Australia, to Columbus, Ohio. On May 9, 1949, the same aircraft commemorated the 1919 flight of the NC-4 by flying from FBF to Lisbon in record-setting time. (TH.)

TRANSCONTINENTAL SPEED RECORD, 1954. On April 1, 1954, three Grumman F9F-8 Cougars of VF-21 (above) set the transcontinental speed record from NAS San Diego to NAS New York. At right, the record-breaking fliers—from left to right, Lt. Wallace Rich, LTJG John Barrow, and Lt. Comdr. Francis Brady— were actually on a routine training flight and did not intend to break the record. They averaged 645 miles per hour, refueling once from an AJ Savage tanker, with Brady making the 2,438-mile trip in the shortest time of three hours, 45 minutes, and 49 seconds. (Both, NA.)

TRANSCONTINENTAL RECORD BROKEN AGAIN. On July 16, 1957, Korean War veteran Marine pilot (as well as future astronaut and US senator) Maj. John Glenn reset the transcontinental air-speed record in a Vought F8U-1P Photo Crusader (above). He and wingman Navy lieutenant commander Charles Demmler (flying an F8U-1) took off from NAS Los Alamitos, but Demmler broke a refueling probe during the first of three air-to-air refuelings (below) and had to abort the flight at Albuquerque, leaving Major Glenn to complete the journey alone. He landed at NAS New York, Floyd Bennett Field, with a flight time of three hours, 23 minutes, and 8.4 seconds. (Above, Vought; below, NNAM.)

UNIQUE PHOTOGRAPHIC RECORD. While breaking the speed record, Major Glenn also recorded the first transcontinental, panoramic photograph of the United States. Glenn's transcontinental flight was given the name "Project Bullet" because his Crusader flew faster than a round from a .45-caliber pistol. That speed is captured in the "self portrait" (above) of Glenn's shadow speeding over the blur of the ground. The final frame from his panorama (below) shows his family, with his wife stepping forward from the crowd, awaiting his arrival at Floyd Bennett Field. Glenn's Crusader continued its Navy career until it was lost in an accident while landing on USS *Oriskany* off the coast of Vietnam on Friday, December 13, 1972. (Both, NA.)

FBF AND THE BENDIX TROPHY, 1961. "Project LANA" (a complex acronym combining the Roman numeral *L*, meaning 50, with Anniversary of Naval Aviation) was planned to break the three-hour transcontinental barrier. Five brand-new McDonnell F4H-1 Phantoms left Ontario Airport (the easternmost of Los Angeles's airports) for NAS New York. Lt. Comdr. Richard Gordon and LTJG Bobbie Young in *LANA 3* (above) have just landed, trailing their drag chute, after setting the record time of two hours and 47 minutes, averaging 903 miles per hour. Note the Manhattan skyline visible in the distance. Below, the winning crew (below) of Gordon (left) and Young admire their Bendix Trophy. Richard Gordon served in the Vietnam War and later became a NASA astronaut, flying on the Gemini 11 orbital and Apollo 12 moon missions. (Both, TH.)

TRAGEDY AND RESCUE. After the war, while NASNY focused on Reserve training, Coast Guard Air Station Brooklyn returned to its main purpose of rescuing those in peril. In September 1946, a Belgian Sabena Airways Douglas DC-4 airliner crashed in a heavily forested area near Gander, Newfoundland. Search planes noted survivors at the inaccessible crash site. The Coast Guard dismantled and airlifted their two remaining operational Sikorsky helicopters, a surviving HNS-1 and a newer HOS-1, from FBF to Gander in a Navy R5D Skymaster (above). A makeshift landing pad (below) had to be constructed in a muddy clearing. While there were 23 fatalities, the 18 survivors were taken by helicopter, one at a time, from the crash site to a makeshift hospital. (Above, NNAM; below, USCG.)

CGAS Brooklyn, 1950s. In the late 1940s and early 1950s, the Coast Guard relied mainly on seaplanes and amphibians rather than the helicopter, which they had pioneered in 1943. The FBF ramp above shows both a Sikorsky HO3S-1G Dragonfly and an example of perhaps the Coast Guard's most significant amphibian, a Grumman UF-1G Albatross. Nicknamed the "Goat," USCG Albatrosses served from 1951 to 1983. A Coast Guard Goat also set numerous speed and altitude records for amphibians. Below, the crew of a Martin P5M-2G Marlin scrambles for a rescue; the Marlin was a derivative of the wartime PBM Mariner. With a crew of 11, it had a top speed of 250 miles per hour and a range of more than 2,000 miles; it only served from 1955 to 1960. (Both, USCG.)

HELICOPTERS REGAIN IMPORTANCE. The Sikorsky HH52A Seaguard, seen above, became the rotary-wing backbone of the Coast Guard in general, and Air Station Brooklyn in particular. It also could routinely operate from the small landing platforms on cutters and icebreakers. Introduced to the USCG in 1963, it served until 1989, saving approximately 15,000 lives. Beginning in 1984, the HH52A was replaced by the Aerospatiale H-65 Dolphin, shown below inside the typically pristine Air Station Brooklyn hangar. With a crew of two pilots plus a mechanic and rescue swimmer, it remains the most numerous short-range rescue helicopter, with 100 still in USCG service. (Above, USCG; below, author.)

NEW YORK AIR NATIONAL GUARD. Floyd Bennett Field was also the home of the New York Air National Guard's 106th Wing, the country's oldest Air National Guard unit, from 1947 to 1970. It gained federal recognition in March 1947, flying Douglas B-26 Invaders. On February 1, 1951, the unit was mobilized. The photograph above shows officers and enlisted guardsmen mustering in front of Hangars 9 and 10, which were originally built for the Navy's Air Ferry Service. The unit's North American B-26 Invader bombers are in the background. The unit converted to the Boeing B-29A Superfortress at March Field, California, pictured below, for the training of reservist crew members for reassignment to combat units flying in the Korean War. (Both, NYANG.)

NYANG B-26 INVADER. The 106th Bombardment Wing was released from federal service and returned to FBF in 1953. Its component 102nd and 114th Bombardment Squadrons resumed flying the B-26B Invader attack bomber. Developed during World War II as the Douglas A-26, the Invader was a fast, powerful light bomber with fighter-like performance. During the Korean War period, the main mission was night interdiction, hence the black paint scheme with subdued markings (above) on this example taxiing at FBF. Flown by a single pilot, the crew included a navigator/bombardier, who sat in the acrylic (Plexiglas) nose compartment (at right), and a gunner. (Both, NYANG.)

CHANGE OF MISSION. In July 1956, the wing was redesignated the 106th Fighter Interceptor Wing, and the component 102nd and 114th Fighter Squadrons were reequipped with Lockheed F-94B Starfires (above). This first-generation jet interceptor was developed from the T-33A trainer. It was equipped with a Pratt & Whitney J-48 turbojet with an afterburner. The pilot sat in the front, while the radar intercept officer sat behind, with his head buried in a radarscope. Its service life was short as better aircraft were introduced. At FBF, it was replaced by the North American F-86H Sabre (below), the ultimate version. of this iconic jet, equipped with a more powerful General Electric J-73 engine and armed with four 20-millimeter cannons. (Both, NYANG.)

FURTHER CHANGES OF MISSION. In 1958, the wing was redesignated again, this time to the 106th Aeromedical Transport Group. The 114th Squadron was disbanded, but the 102nd Squadron transitioned to Fairchild MC-119J Flying Boxcar cargo aircraft, seen above. Designed specifically for the medical evacuation role, it could carry litter patients as well as attending medical staff. In 1963, the mission and aircraft changed again, this time to the 106th Heavy Air Transport Wing, flying the Boeing C-97A/G Stratofreighter. The image below shows a C-97 undergoing maintenance in the "nose" hangar (the name implies that the tail cannot fit in) built for the ANG adjacent to the administration building. (Both, NYANG.)

FINAL CHANGE OF MISSION AND THEN STATION. In September 1969, Floyd Bennett Field's ANG wing was redesignated yet again, this time becoming the 106th Air Refueling Wing. The 102nd Squadron's new aircraft was the Boeing KC-97L Stratotanker (above); however, the wing's time in Brooklyn was coming to an end, as the Navy decided to close NAS New York. The 106th Wing transferred about 90 miles east to the former Suffolk County Air Force Base, today known as the Francis Gabreski Air National Guard Base. It is currently the 106th Air Rescue Wing, flying the Lockheed HC-130P Hercules and Sikorsky HH-60G (below), as one of only three combat search-and-rescue units in the Air National Guard. (Both, NYANG.)

Five

NAVAL AIR STATION TO RECREATION AREA

FLOYD BENNETT AIRPORT, TODAY. After the Navy left, the airfield was turned over the National Park Service. The view from Flatbush Avenue has not changed significantly since the early 1930s, including the signage. The seal of New York City appears below the name, showing a colonist and Native American holding a shield with beavers and flour barrels—symbolic of the commercial heritage of the city. (Author.)

ADMINISTRATION BUILDING. The photograph above shows the "land side" of the administration building with the passenger-entrance portico; the lower-level baggage entrances are partially hidden by bushes. Visible in the background, the so-called "nose" hangar built for the NYANG has since been demolished. The "air side" view at left shows the passenger-access stairs to the ramp on each side of the control tower projection. Compared with the 1930s photographs, the Navy's modernization of the control tower is readily apparent. Note the building's right entablature is unchanged. On the left entablature, beyond the view in this photograph, "Naval Air Station" replaced the title "City of New York" in 1941. (Both, author.)

ART DECO INTERIOR. While its exterior is a mixture of styles, the administration building's interior is purely Art Deco. Unlike other airports, this one building combined the functions of airport offices and the passenger terminal. Passengers entering through the portico were greeted by an airy first-floor lobby (at right). Offices and shops were located off hallways to the right and left. A balcony (below) provided a view of the lobby and stained-glass ceiling as well as access to more shops, offices, and a restaurant. A WPA-sponsored Federal Art Project mural was taken down from the lobby in the late 1930s after complaints it was "communist inspired." (Both, author.)

ADMINISTRATION BUILDING AND HANGAR ROW. Looking south, Hangars 3 and 4 can be seen beyond the administration building (above). Two of the tunnel exits onto the ramp are just discernable on the pavement. A closer view of Hangar 3 (below) shows traces of the original "City of New York Department of Docks" title. The assignment of the airport to this department gives some indication of the city's mind-set at the time, with a focus on New York City as a seaport. As originally built, each pair of hangars was separate and freestanding. To gain workspace and offices, "lean-to" brick structures were added at each end, as shown here. (Both, author.)

HANGAR CONSTRUCTION. In 1936, to gain additional workspace and offices for the naval reserve air base and other tenants, each pair of the freestanding hangars was joined by a central structure (shown here). The brick additions matched the overall structural design and were finely detailed with a curved parapet featuring a stylized wing and propeller and the letters "NYC" and "FBA" (for Floyd Bennett Airport). (Author.)

COAST GUARD STATION, NOW NYPD. CGAS Brooklyn closed in 1998, and the New York City Police Department took over its base. The NYPD formed its aviation unit in 1919. Based at nearby Sheepshead Bay from 1928, it moved to FBF's Hangar 4 sometime after the airport opened. The NYPD has been a tenant ever since, and its helicopters form the sole remaining flying unit at FBF. (Author.)

HANGAR B. In response to President Roosevelt's 1939 declaration of a "Limited National Emergency," the Navy began expansion of facilities at FBF and elsewhere. The first step was leasing more land and building a Neutrality Patrol base around a new Hangar A on the eastern side of the airfield. After entry into the war, Hangar B (above) was added. Today, of these two, only Hangar B remains, serving as the home of the Historic Aircraft Restoration Project (HARP). Staffed by a sadly dwindling group of volunteer veterans, the group is dedicated to restoring or re-creating aircraft significant to FBF's history, like the hand-built Lockheed Vega (below), finished to resemble Wiley Post's *Winnie Mae*. (Both, author.)

LOCKHEED NEPTUNE.
One of the last aircraft flown by the Navy Reserve units at FBF was this Neptune SP-2E, beautifully restored to almost flying condition by HARP. Originally a P2V-5FS, the designation of all Navy aircraft was changed in 1962 to align with the Air Force system because secretary of defense Robert McNamara found the existing system too confusing. (Author.)

NYPD GRUMMAN GOOSE. This 1945 Grumman G-21A Goose has also been beautifully restored by HARP; it is in the colors used by the NYC Police Department Aviation Unit. Although today an all-helicopter force, the unit flew seaplanes and amphibians from its founding to 1954. Its position as the last photograph in this book is appropriate since the NYPD is the last remaining flying unit based at Floyd Bennett Field. (Author.)

DISCOVER THOUSANDS OF LOCAL HISTORY BOOKS FEATURING MILLIONS OF VINTAGE IMAGES

Arcadia Publishing, the leading local history publisher in the United States, is committed to making history accessible and meaningful through publishing books that celebrate and preserve the heritage of America's people and places.

Find more books like this at
www.arcadiapublishing.com

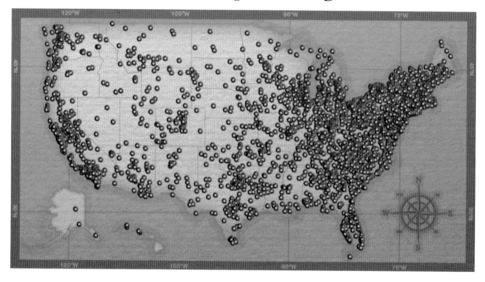

Search for your hometown history, your old stomping grounds, and even your favorite sports team.